Truth

TRUTH

Albert Soriano

ALEXANDRIA
LIBRARY
PUBLISHING HOUSE
MIAMI

Truth
@ Albert Soriano, 2016

ISBN: 978-1540345691

Edition, Composition, Interior and Cover Design:
Vilma Cebrian
Alexandria Library Publishing House
www.alexlib.com

Contents

PREFACE

Truth is immortal.

God's truth surged at the same instant of the universe creation, when physical and natural laws were to rule forever the evolution of all reality.

Billions of years later life and human existence appeared on earth, following the transformation of energy contained in all the matter of the universe.

Since the beginning of civilizations, brilliant minds reflected on these events with some of them expressing the followings:

Truth is reality.

Truth is a fundamental characteristic of God.

Genesis is the book of the beginning; it is the place where we find the word Truth together with the beginning of man.

Plato said:
"God is truth"
"Truth is Divine"

Chapter I
LIFE

The first life in earth was microbial which for billions of years evolved in the oceans, creating many complex forms and eventually many of them moved to the dry lands to keep evolving into a vast array of animal species that thrived on earth's oceans, lands and atmosphere.

Life is the result of the same physical and natural laws that originated the universe, at the instants of the Big Bang, when matter with active energy, in space-time was shaped as gigantic clouds of dust and gas, called Nebulas, the cradle of all stars that emitted light and heat in Billions of galaxies at every corner of visible Universe.

The visible universe with active energy is just 5%, which together with 95% of CDM, cold dark matter, with energy in state of rest encompasses the whole universe.

CDM makes possible the law of gravity, to bind both energies, the active and the one in state of rest, in the universe.

Gravity is the most important physical law that made possible life on earth and by continuing evolving for millions of years, eventually created the Homo Erectus a bipedal animal, able to walk with two legs and by freeing its 2 hands, accelerated the expansion of its brain mass, to process thought and reasoning of all the reality of the natural environment surrounding the precursor of the human species.

In addition, the inborn natural instincts, were to be conveyed to primeval human thru the generational chain giving to them the hability to group and to act, first in the small groups called tribes, increasing their chances of survival, and by continuing along this path, societies were finally organized, with larger numbers of individuals within abstract boundaries defining their territory in the natural environment, where each member was to be safer, allowing them to increase their efficiency in performing assigned tasks according to their skills.

Nevertheless, the individual was to keep their own identity but subjected to certain responsibilities that made possible a peaceful coexistence, where ideas were to be shared favoring the whole group during the struggle for survival and their adaptation to natural habitats.

It was then that humans started to make tools from stone and metals, used for hunting and other activities.

Those times have been described as the Stone and Bronze Ages, which set the foundation for larger societies to be reorganized as civilizations all over the world.

Meanwhile, natural events like conception, birth and death, due to natural or violent causes, were to be recognized as inexorable realities faced by each generation during their struggle for survival. Migrations of humans commenced instinctively al the same time and was to become a major step for humans in seeking thru explorations of new natural habitats, which could offer better chances to survive. Migrations stills today is a common practice for the relocation of larger masses of people, replicating physical and natural laws, that always produce positive, beneficial results, when humans escape from negative events like warfare, enslavement and famines, which disrupt the peaceful equilibrium required in any society.

It is interesting that humans exist today because of the gradual warming of Earth, after its latest glacial age when an ice sheet of many miles deep, covered the northern hemisphere of our planet, where Europe, Asia and North America are located today.

Because of that warming, Homo Erectus started to migrate North into Europe from Africa, seeking a new habitat, opening to them, with abundant animal life that could provide for sustenance and any other survival need.

This is why the new species of Homo Neanderthal was to be recognizable in Europe with new white skin color and for their printings in the walls of many caves depicting the animal life that roamed Europe at that time.

Glacial ages are described by scientists as a phenomena caused by the Earth's orbit around the Sun, which also follow its own path around the center of our Milky Way galaxy in the Universe.

It is estimated that the sun's orbit, carrying Earth, and other planets, last 225-250 million years.

The Earth currently is in an interglacial period (glacial retreat) called the Halocene that could last 16,000 years.

In early times, Humans adapted to their natural environment, ruled by Physical and natural laws, where primitive humans learned how to make fire when observing the lighting creating fire in the forest, and to replicate it by striking 2 stones to create an spark to make fire, that was to warm their habitats in caves and to cook their raw meals.

Much later on, civilizations made possible, thru science and technology, the understanding of all physical and natural laws, applying them to improve their way of life in all aspects of production, transportation, etc; since the truthfulness of those laws always was to produce positive deterministic outcomes, that eliminate all negative influences in any society, such as false ideologies and even more important, the no practicing of truth, jeopardizing human coexistence.

Also, at that time, Humans were instinctively developing a belief in a superior being that was directing all the events around them in the natural environment and on the heavens.

Today humans have scientifically proven that all physical and natural laws explains the vast array of events in the reality of all, including the presence of Humans on earth, a planet orbiting the Sun in the vastness of the Milky Way Galaxy, one among millions of galaxies in the Visible Universe.

Nevertheless, after all the scientific discoveries throughout many civilizations, humans are in accord that God, a superior being, is the Creator of all Reality, and since humans are naturally humble, they also understand that the only thing that mankind will never know is what God had in mind when created all the reality of the universe and Life on Earth.

"I think therefore I am"

From this, the existence of God was confirmed. Descartes said that:

"God's existence is proven by:

1.- Everything have a cause.
2.- The fact that my idea of God represent an infinite perfect being, must have cause, which is infinitely perfect.
3.- I am not infinitely perfect."

Rene Descartes
17 th Century French Philosopher

Thousands of years earlier, it was written in the Torah, Haftorah Ha Azinu (page 1255)

"For all his judgments were before me
and I did not depart from his statues"

From those words humans must accept to believe that only thru the practice of Truth, like those contained in the physical and natural laws, Humans will be able to continue coexisting on Earth since they "did not depart from God's statues"

"To tell us that every species of thing is endowed with an occult specific quality by which it acts and produces manifest effects, is to tell us nothing; but to derive two or three principles of motion from phenomena, and afterward to tell us how the properties and actions of all corporeal things follow from those manifest principles, would be a very great step"

Isaac Newton
OPTICS

Chapter II
MATTER AND ENERGY

Matter and Energy are the fundamental entities in the Universe that exists since their creation in space–time.

Atoms last forever and contain active energy, creating 5 % of all the matter in the visible Universe.

In contrast, CDM, cold dark matter, with energy at rest, represents the other 95 % of the whole Universe.

Matter is the substance from which all materials are made.

Everything in the Universe is made of matter and energy and includes Humans on Earth.

The most important physical laws in the Universe are:

> 1.- Matter
> 2.- Energy
> 3.- Elements
> 4.- Light
> 5.- Gravity
> 6.- Electromagnetism

1.- Matter:

Can be solid, gas or liquid.

Matter in solids, its atoms or molecules have no freedom or mobility; they only vibrate in their fixed positions.

Molecular dynamic is the new science that explains the structural geometrical forms of all matter in the Universe.

Matter states are:
Solid
Liquid
Gas
Plasma

2.- Energy

It is a property of matter. It can be transferred between objects and converted into forms.

Energy cannot be created or destroyed.

There are two types of energy: Potential (stored energy) and Kinetic (energy in use)

Energy includes light, nuclear electrical and chemical energy.

Cold dark matter CDM accounts for 95% of the Cosmos, with energy in state of rest. The remaining 5 % is the visible universe with active energy represented by nebulas, galaxies, stars, planets, and many other celestial bodies, but, there is still pending the discovery of a multiverse structure, where 2 or mores universes are connected to each other.

Black holes exist in the Universe, at the center of each galaxy, like our own Milky Way Galaxy.

Black Holes practically swallows stars and any other celestial body, which disappear from view, to probably confirm the theory that Black Holes are the conduits between Universes to integrate a multiverse structure.

3.- Elements
See Chapter III

4.- Light
It is another fundamental physical law that originated the visible universe, where our sun exists in one of the arms of the Milky Way Galaxy The Sun's light rays are a particle and wave that takes 8 minutes to reach the Earth to create and support life and its evolutionary progress, making possible the presence of the Human Species and unaccountable other species in our planet.

5.- Gravity
Is a natural Phenomena by which all things with mass are brought toward (of gravitate toward) one another including planets, stars and galaxies.

On Earth gravity gives weight to physical objects and causes the ocean tides.

6 - Electromagnetism
The Electromagnetism force is the one responsible for practically all the phenomena one encounters in life above the nuclear scale, with the exception of gravity

When humans "push" or "pull" ordinary material objects, this result from the intermolecular forces that act between the individual molecules in our bodies and those in the objects.

"God is able to create particles of matters of several sizes and figures.and perhaps of different densities and forces and thereby to vary the laws of nature, and make worlds of several sorts in several parts of the Universe. At least, I see nothing of contradiction in all this"

Isaac Newton
OPTICS

"To the entire world added our father the Sun, I give my light and my radiance; I give men warmth when they are cold; I cause their fields to fructify and their cattle to multiply; each day that passes I go around the world to secure a better knowledge of men's needs and to satisfy those needs. Follow my example."

An Inca Myth recorded in the "Royal Commentaries" Of Garcilasso de la Vega.1556

Albert Einstein famous Equation: $E = mc^2$
E= Energy of a physical system
M= is the mass of the system
C=The speed of light (in a vacuum)
Or Energy equals mass multiplied by the speed of light squared, this implies that any small amount of matter contains a very larger amount of energy.
Albert Einstein
General theory of relativity. 1921

Chapter III
ELEMENTS

Elements are the substance that cannot be broken down farther. Natural elements are result of the Big Bang that created the Universe. Elements are numbered from #1 to # 92, being Hydrogen, #1, the most abundant element in the Universe, and # 92, which is uranium, from where Humans discovered, in the last century, that Uranium could be transformed into atomic energy for peaceful purposes like electricity or for weapons of warfare, like the atomic bomb, which was used for the first time in World War II, when the U.S dropped the atomic bomb over Japan that ended the war with the U.S. in 1945.

Elements exist in every life form on Earth.

The body structure of Humans is 62 % water and 16 % Protein, with the total elements consisting of:

	%
Oxygen	65
Carbon	18
Hydrogen	10
Nitrogen	3.5
Calcium	1.5

Phosphorus	1
Potassium	0.35
Sulphur	0.25
Sodium	0.15
Magnesium	0.05
Iron	0.70

And traces of copper, zinc, selenium, lithium, aluminum, silicone, lead, arsenic, bromine.

Elements are unique atomic structures with diverse number of protons, electrons and neutrons, all swirling around the atom center, its nucleus, to represent all matter in the universe and all life on Earth.

Humans are a complex and sophisticated structure of elements that surged initially as a microbe on Earth to keep evolving for billions of years, till the primitive human Homo Erectus, appeared in the African Continent.

Thousands of years of evolution transformed Homo Erectus into a new species, known as Homo Sapiens, that was to revolutionize the survival scheme by replacing the animal force with machines in the past few centuries.

Humans kept building machines of all types, advancing even farther, in the last century, with robotic machinery that today and in the near future, could even replace completely Humans in the manufacturing of goods, transportation and most activities in our Civilization.

Recently, with the discovery of the transistor, that made possible the computer and all electronic equipments and gadgets, like the internet, cell phones, etc, with all of them using a new electronic language the one of 0 and 1, a binary electrical system, moving at the speed of light of 186.000 miles a second, have literally shrank our Planet thru a worldwide communication, and even in Space, with satellites circling the Earth and in Space exploration by traveling in the Solar System and beyond in the near future using the new clean energy from Hydrogen, to power spaceships carrying humans into the Milky Way Galaxy.

"A Community of matter appears to exist through the Visible Universe, for the stars contain many of the elements which exist in the Sun and Earth.
It is remarkable that the elements most widely diffused through the host of stars are some of those closely connected with the living organism of our globe, including hydrogen, sodium, magnesium and iron.
May it not be that at least the brighter stars are like our Sun, the upholding and emerging centers of systems of worlds, adapted to be the adobe of living beings"

William Hugging
1865

"We do not ask for what useful purpose the birds do sing, for songs is their pleasure since they were created for singing.
Similarly, we ought not to ask the human mind troubles to fathom the secrets of the Heavens.
The diversity of the phenomena of nature is so great and the treasure hidden in the Heavens so rich, precisely in

order that the human mind shall never be lacking in fresh nourishment "

Jonathan Kepler
Mysterium,
Cosmo Graphicum

Chapter IV
HUMAN BEHAVIOR

At this juncture in human history there is only one alternative left for our species continued presence on Earth and it is to develop a positive behavior, only obtainable thru the practice of Truth, at every instant of human coexistence.

Truth will be acting as the Human Equalizer to all Physical and Natural Laws, which precisely carry also an inalterable Truthfulness because of their positive deterministic outcomes.

Today, Humans understand that to reach the main good of survival, only this could be possible by the elimination of the rampant carnage of our species during the past few centuries that easily could escalate into a total destruction of the present Civilization, provoked by the use of atomic weapons in a worldwide conflagration.

Human behavior is the main factor that must be recognized once it surges at the moment of a new born, that while gulping its first breath of air, instinctively cry, by recognizing his entering into a new hostile environment, in contrast to the one of the mother's womb that protected and fed its new life.

It is precisely at the moment of birth, that every human develop its first negative behavioral reaction, when crying, sensing the new environment to be full of hardships and without any joy.

At this point, parents, family and society in general, must care for this new life, by observing closely while growing up, the child reactions and inclinations, but above all is to encourage his curiosity, the most important instinct in humans, that when reaching adulthood, many of them will become, explorers, scientists, inventors, philosophers, writers and technologists, the core from which later on, in a few centuries, that society will be recognized as a Civilization.

Otherwise, negative influences and false ideologies, present in any Society, will create an antisocial behavior, where civilizations such as the Roman Empire, which after many centuries of being the most powerful nation in the world at that time, finally joined the dust of History.

This event is symbolized in the prophecy of the "Four Horsemen of the Apocalypse"

1. *Antichrist*
2. *Wars*
3. *Famines*
4. *Death*

In our times, theses could be translated as:

1. *Lying*
2. *Deceit*
3. *Corruption*
4. *Vices*

There is another subtle statistic that reflects the underlying reality of Human species fearing its extinction, and it is the skyrocketing increase of its population, which already have reached the 7 Billion mark and growing at the rate of 1 Billion almost every 10 years, since the beginning of the 20th Century and projected to top the 10 Billion by 2035.

All these numbers are in stark contrast to the human population growth from the first humans on Earth till the year 1900 when the world population numbered only 2 Billion.

It is reasonable to assume that the present extraordinary growth of world population is because of our Civilization advances in sciences and technology during the past 2 centuries.

But we could also explain the why of such a rapid growth in the last centuries because humans instinctively fear a devastating conflagration, where weapons of mass destruction, like the atomic and hydrogen bombs, could be used and bring about the extinction of the human species.

In contrast there are natural phenomenas like the Tectonic Plates moving slowly (few centimeters each year) all over the World's crust rearranging the existing continents, which 230 millions of years ago, was only a single continent called PANGEA.

Scientists predict that today's continents geographic locations will eventually converge once again to became a single continent, millions of years from today.

As an example, scientists predict that Australia corresponding Tectonic Plate, will keep moving North, for in 1 or 2 millions of years, the Australia continent will be located where the Philippines are today in the Pacific Ocean, thousands of miles North.

Mankind's only choice will be adapt and delineate new geographical and political frontiers, a task to be carried out during the evolution of thousands of future generations of humans.

Another probable natural event might be the one of an asteroid or other celestial body impacting the Earth, to cause enough damage to its peoples and cities that they could even recover from, but never like the one that 65 millions of years ago, caused the extinction of the dinosaurs and most living species on Earth.

Chapter V
TRUTH

The Physical and Natural laws of the Universe only creates positive deterministic outcomes.

Deterministic is a general philosophical thesis that for everything that happens, there are conditions such that given them, nothing else could happen.

Truth is deterministic in nature, and always reaches its final objective, or reveals itself sooner or later, by a given set of initial conditions.

The best example is the prescence of the Human species on Earth, the third planet from the Sun, one among billions of stars in the Milky Way Galaxy.

It took many Eons for the Human species to develop its most important characteristic, not seen in another life species, and it is the capacity to think and to reason, which in addiction to natural instincts, passed along the generational chain, allowed the primeval Humans to confront their survival in the Natural Environment as an individual and later on as a social being, with spoken and written language to communicate with members of their own species around them and beyond.

One word, Truth, was to become the cornerstone of human integration as a Society, all over the world.

It is significative that Truth always carry positive deterministic outcomes to every human action during the survival struggle, because Humans instinctively follow the Physical and Natural Laws that with their inexorability are able to maintain a positive, livable balance in the natural environment that offers Humans their means for survival.

Humans know today that the Reality of all life and the Universe, both seemingly chaotic, are instead harmonious and complex systems, organized by matter and energy while being transferred forever in Space-Time.

Therefore, Truth acting as an Equalizer, binding Humans to all Physical and Natural Laws, like seasons for humans when to plant and to harvest crops, to follow animal migrations for hunting and obtain nourishment, etc.

Much later, Humans learned how to use the river currents and wind, to generate power to run their first machines to process grains.

Centuries later rivers were to be tamed to run turbines generating electricity, the new discovered power to move machines and that lighted their cities and homes.

At the same time fossil fuels were joining the sources for energy, like those from coal, oil and natural gas, revolutionizing the output of goods production, during and after the Industrial Revolution, when machines were to be used also for transportation on land, rivers and oceans, and soon, into the air.

But, the 20th Century was to be the Era of the atomic energy, obtained from uranium, # 92 in the table of elements, providing an inexaustable source of energy. Unfortunately this energy was also to be used in warfare, as the atomic bomb, a weapon of mass destruction.

But, science soon will provide a similar inexaustable source of power and it is from Hydrogen, the most abundant element in the Universe which by not being radioactive could replace the atomic energy from Uranium, since Hydrogen, besides creating electrical power could also be used by space ships carrying Humans in the exploration of the Solar System and beyond in our own Milky Way Galaxy, in not a distant future.

Finally, the author would like to mention a unique event in our Civilization History where the significance of the word Truth is clearly reflected, without any doubt, when in the creation of a new Nation. The United States of America, in its Declaration of Independence on 7/4/1776, the American colonies proclaimed their fundamental reasons in seeking freedom from the British Empire, by saying:

"That the Laws of Nature and of Nature's God entitle its people to separate from the British Empire"

Adding:

"We hold these Truths to be self evident and that all men created equal and endowed by their Creator with certain unalienable rights, that among them are life, liberty and the pursue of Happiness"

Just by reading the above outlined excerpts of the Declaration of Independence, those patriotic humans were in reality aligning their Free Individual existence to those Physical and Natural laws ruling the Reality of all in the Universe.

<u>Excerpts:</u>

1.- Laws of nature
2.- Nature's God
3.- Truth
4.- Men are created equal and endowed by their creator
5.- Certain Unalienable Rights
6.- Life, Liberty and the pursue of Happiness

Do you know the ordinances of Heavens?
Can you establish their role on Earth?

The Book of Job

It is not from space that I must seek my dignity, but
from the government of my thought.
I shall have no more if I possess worlds.
By space, the Universe encompasses and swallows
me like an atom; by thought I comprehend the world.

Blaise Pascal
Pensees

Chapter VI
REALITY AND TRUTH

Reality is the conjectured state of the thing as they actually exist, rather than as they may appear or might be imagined. Reality includes everything that is and has been, whether it is observable or comprehensible.

A broad definition of Reality is everything that has existed or will exist.

Reality is the higher truths than less abstract concepts so truth will always refers to what is real.

Reality is in the totality of all thing, structures (actual and conceptual), events (past and present) and phenomena, whether observable or not. It is what a world view (whether it is based on individual or shared human experience), ultimately attempts to describe or map the Truth and refers to what is Real.

Truth means accord with fact or reality or fidelity to an original or to a standard or ideal.

"We look back through countless millions of years and see the great will to live struggling out of the intertidal slime, struggling from shape to shape and from power to power, crawling and then walking confidently upon the land, struggling generation after generation to master the air, creeping down into the darkness of the deep; we see it turn upon itself in rage

and hunger and reshape itself anew. We watch it draw nearer and more akin to us, expanding, elaborating itself, pursuing its relentless inconceivable purpose, until at last it reaches us and its being beats through our brains and arteries. It is Possible to believe that all the past is but the beginning of a beginning and that all that is and has been is but the twilight of the dawn. It is possible to believe that all that the human mind has ever accomplished is but the dream before the awakening... Out of our... lineage, minds will spring, that will reach back to us in our littleness to know us better than we know ourselves. A day will come, one day in the unending succession of days, when beings, beings, who are now latent in our thoughts and hidden in our loins, shall stand upon this each upon a footstool, and shall laugh amidst the stars."

H.G.Wells, "The Discovery of the futures"
Nature 65, 326 (1902)

"To what purpose should I trouble myself in searching out the secrets of the stars, having death or slavery continually before my eyes?"

A question put to Pythagoras by Ansximenes
(c. 600 B.C.) according to Montaigne

Truth in Practice
Truth is for grabs
Truth is the ultimate meaning of our lives
Truth is the solid rock on which we can build our lives.
We live in an stormy world.

Richard Steams
SPU-Response. Summer 2013

Chapter VII
SYNCHRONOLOGY

Years	Reality	Truth
13-8 Billion	Universe creation (Big Bang)	
12-13 Billion	Milky Way Galaxy (150,000-180,000 Light Years Diameter)	
4.54 Billion	The Earth was formed	
3.8 Billion	Single cell organisms (Bacteria) commenced its evolution	
2.8 Billion	Multicellar life	
400 Million	Macroevolution variations in animal species on the oceans and earth, all with 21,000 genes as a common denominator	
230 Million	Fragmentation of the super continent Pangea. Volcanic activity.	
180-100 Million	Dinosaurs appear on Earth	
65 Million	Dinosaurs and most species extinct by a Meteorite or large comet impacting the Earth. Small placental animals survive, for later becoming the precursors of the human race	
5-10 Million	Apes started to evolve into Hominides	
2.8-1.5 Million	Homo Habilis handyman-Tool use	
1.9 Million - 70,000	Homo Erectus / Primaveal Humans	

Years	Reality	Truth
50,000	Homo Neanderthal. Great hunter, buried their dead. Europe and Middle East.	
50,000-30,000	Homo Sapiens. Cave men. Europe (paintings). Gutural sounds into rudimentary linguistic skills.	
15,000	End of Last Glacial Age	
14,000	Humans Migrate	
5,000 BC	Human used hieroglyphics to communicate in writing (Egypt)	
3,000 BC	Chinese and Hebrew civilizations create writings as we know it today.	
3,000 BC - 30 BC	Pharao Dynasties	
1,200 - 800 BC	Greek City States. Death of Alexander the Great.	
424-348 BC		Plato: "Humans only see the shadow of Reality. Truth is the beginning of every good in Man. Dictatorships arise out of Democracy, Tyranny out of extreme liberty"
400 BC		Democritus: "Matter is made up from atoms"
1-33 AD	Life of Christ. 200 Million People on earth	
60-120 AD		Epictetus. Stoicism must be practical. If you find the Truth you become invincible.
30 BC - 284 AD	Roman Empire	
500 - 1400 AD	The Dark Ages	
1135 - 1204 AD		Maimonedes: "It is impossible for the Truth arrived by Human intellect contradict that revealed by God."

Years	Reality	Truth
1452 - 1519	Renaissance (14th - 17th Centuries)	
1492	Columbus discovers America	
1632		Baruch Spinoza: "Everything that existing in Nature / Universe is one Reality (substance) that has the properties of both extension and thought. You are apart of God as your heart is apart of your inmortality through scientific knowledge".
1776	American War of Independence	
1789	French Revolution	
1800	1 Billion people inhabit the Earth / Industrial Revolution	
1850	Electric Motor, light, telephone, telegraphy, automobile	
1890	Motion pictures, airplanes.	
1900	Spanish - American war. USA becomes a World Power	
1914 - 1918	World War I	
1930	Two billion people on Earth	
1933		Einstein's correpondence with Freud: "Apolitical ways of education could remove the force withtin humans of a wish to hate and destroy."
1938 - 1945	World War II. Germany Italy and Japan defeated. Atomic bombs in Japan.	
1939 - 1945	Holocaust in Europe. 6 Million Jews murdered in gas chambers by Nazi Germany.	
1960 - 1977	Satellites, Space flight, Genes mechanism, Internet, Nano technology	
1961	Korean War. 3 billion people inhabit the Earth.	
1955 - 1975	Vietnam War	
2000 - 2001	New Century, Twin Towers down in New York.	
2008	Jewish New Year 5769.	

Years	Reality	Truth
2009	6 billion people on Earth	Michael Shermer: "Is religion good for social health?" "The fundamental social unit in our evolutionary history arose long before religion and government."
2016	7 billion people inhabit Earth	
2024	8 billion people on Earth (estimate)	
2045	9 billion people on Earth (estimate)	

Comment

You can estimate, from your date of birth, how many generations have lived on Earth before you.

The average time for a human generation is approximately 25 years, which represents 4 generations in every 100 years.

> *If you were born in the years:*
> *1950 equals 4 generations from 1850*
> *1950 equals 8 generations from 1750*
> *1950 equals 12 generations from 1650*
> *And so on and so on...*

Or 40,000 generations in 1 million years since Homo Erectus first presence on Earth. Doesn't make you think when reading "Synchronology" that you are traveling into the past in a time-machine?

ABOUT THE AUTHOR

Albert Soriano was born in Santiago e Cuba, province Oriente, Cuba in 1928. His family moved to Havana at time of the 1933 Revolution. In 1954 Mr. Soriano graduated from the University of Havana with a degree in Accountancy while working for a subsidiary of Procter and Gamble, an American corporation doing business in Cuba at that time. In 1955 he established a Certified Public Accountant firm in Havana.

In 1960 all private businesses in Cuba were confiscated by a communist revolutionary regimen which eliminated private ownership. In 1961 Albert Soriano went into exile with his wife and children to Miami, Florida, where they still reside today.

Mr. Soriano has several essays on History and Science, and have published works such like. "*A Human Self-Reflection of Social Evolution*" (Miami, 2012) and "*For a Peaceful Mankind*" (Miami, 2016).

In this new work "*Truth*", Albert Soriano once again explores his interest on life, civilizations and the future of mankind.